ABOVE: *The village cooper at Carlton Cooperage in Bedfordshire.*
COVER: *Ray Cotterell, Master Cooper, at work in his cooperage at the Museum of Cider, Hereford.*
OPPOSITE: *A nineteenth-century village cooper, William Dalley of Turvey, 'plough-wright, cooper and carpenter, drawn from life April 1830'.*

THE VILLAGE COOPER

K. Kilby

EASTERN

CONTENTS

Printed in Great Britain by City Print (Milton Keynes) Ltd, Denbigh Hall, Bletchley, Bucks.

Sketches from Pyne's 'Microcosm' show coopers at work around 1800.

Repaired brewers' casks. A village cooper would often travel to the nearby small breweries and repair their casks before the busy season. Very small breweries would not employ their own cooper since they would not have enough work for him.

INTRODUCTION

If you were asked to name the greatest invention of all time you would probably say the wheel, or the railways, or the aeroplane, or perhaps one of a number of other inventions which might warrant this distinction. Few if any would consider the humble barrel. Yet for thousands of years most commodities were shipped, or moved, or kept, in barrels. Why? Because barrels were exceptionally strong, with hoops binding the joints into a form of double arch; because they were in themselves a wheel, a means of movement at a time when power was dependent upon the muscles of man or beast; and because certain goods actually benefited by being in a barrel. Without the barrel most goods would have remained right where they were made, or not have been made at all. Few inventions have stimulated such enormous, widespread demand over so many centuries.

This iron-age invention was so workaday and taken for granted that once modern technology caused it to be largely superseded it seems to have been dropped from memory as well as from use. Yet it still persists in the whisky industry, for brandy, for better-quality wines and in a small way in other industries, and some people think that the replacement of the beer barrel by metal containers has not been to the customer's benefit.

These barrels were made in thousands of small coopers' shops dotted all over the country, where coopers and the sons and grandsons of coopers toiled away at their blocks. There are few jobs as physically demanding as coopering, amidst the noise of hammering and the smoke of firing. Few jobs require as much skill as the making of a cask. Coopers serve a seven-year apprenticeship, swinging axes and adzes, drawing long knives and slamming hammers, eventually with such precision that they could guarantee to make a watertight vessel in any one of a surprising variety of sizes and shapes.

Until the middle of the nineteenth century village coopers made all manner of casks, tubs, vats, buckets and bowls, which were essential items of equipment in every home throughout the land. Today what remains of this once vital domestic trade is reduced to little more than a few coopers making tubs for plants, ornamental work and furniture, prized more for its 'olde-world' charm than for its utility.

The wooden bucket. Before the nineteenth century it would normally have had wooden hoops.

Small household vessels made by the village cooper.

THE HISTORY OF
COOPERING

Tomb paintings indicate that straight-sided wooden buckets bound with wooden hoops were made in Egypt as early as 2690 BC, and vessels described as barrels are mentioned in the Bible and by the Greek historian Herodotus writing in the fifth century BC. The Greeks and early Romans used clay vessels called amphorae but the broken potsherds from these vessels become scarce for the period from the second century AD, suggesting their replacement by wooden casks. Pliny wrote that this process began in Cisalpine Gaul. From these beginnings coopering developed in almost direct proportion to the growth of trade.

In medieval times, when a couple were married they would have to pay a visit to the cooper; someone perhaps with a name like Walter le Cuver, soon to be anglicised into the surname Cooper. They needed buckets for water and for milking, bowls for washing and dolly tubs for laundering; they needed pickling vats for the kitchen, storage vats for flour and barley, a churn for making butter, tubs for cheese, and a vat, casks and tubs for home-brewed ale. Many of these casks would last a life-time and become valued possessions.

Some coopers began to specialise in what became known as dry bobbing, the making of cheap frail casks for fruit, fish, shoes, nails, paint, grease, seeds, pearl-barley, meats, butter, putty and money; later were added potatoes, tobacco, sugar, syrup, cork and crockery. Stronger casks were needed for gunpowder, and 'tight' leakproof casks for oils and vinegar. A

A cooper-made fruit bowl.

versatile village cooper would be able to supply all these types of cask.

In the cities coopers came together in guilds, primarily in order to create a 'closed shop', as in Coventry: (a cooper) 'shall not occupie any shop within this citie oneless he agre with the Cowpers of this Citie'. Upon completing an apprenticeship he became an enfranchised freeman of the city. Those who fell foul of the guild had no option but to leave the city if they wished to continue in the trade, and to set themselves up as village coopers.

The expansion of trade under the Tudors stimulated demand for the cooper's craft, but even during the sixteenth century developments were taking place which were to have a profound effect on coopering. Breweries were being established and growing fast: by 1591 there were 'Twenty great brewhouses between Fleet Street and St Catherines' and they tended to attract many of the best coopers. The making of beer, wine and spirit casks requires far more expertise and these 'wet' coopers were offered higher remuneration and some relief from the monopoly of the guilds. The Coopers' Company fought to maintain the independence of this branch

of coopering and insisted that brewers bought their casks from independent cooperages. They took their case to Parliament, bribed the Lord Chancellor with half a butt of malmsey in 1533, the Lord Chief Justice with twenty gallons of sherry in 1561 and the Speaker with a runlet of ten gallons of sack in 1562, but although they managed to get Parliament to pass laws in their favour the brewers flagrantly disregarded them. Perhaps they paid bigger bribes. So well did these breweries proliferate that in 1751, when Thrales brewery was being sold to Barcley, Dr Johnson said: 'Sir, we are not here to sell a parcel of boilers and vats, but the potentiality of growing rich beyond the dreams of avarice.'

Although the larger breweries were employing hundreds of coopers up until the 1950s, independent cooperages still produced casks for brewers. The biggest of the independent cooperages was Shooters, Chippingdale and Colliers of East London, which employed 630 coopers in 1900. In fact competition from other means of holding liquids forced the larger cooperages to rely more and more upon the breweries. Wilsons of Bermondsey turned

6

ABOVE LEFT: *A bever barrel, taken by farmworkers into the fields for their dinner break. It would hold from two pints to a gallon of beer.*
ABOVE RIGHT: *A barrel chair was an item of furniture common to the poorer homes.*

from making water and sperm-oil casks to beer casks at this time.

These larger concerns made other efforts to come to terms with changing demands by equipping their cooperages with machinery, but the expense of installing a considerable number of machines to duplicate the wide variety of coopering skills and the fact that the increased output came at a time of diminishing demand made their operations less viable. It was still practical for Samuel Kilby's of Banbury to continue making casks by hand until the 1950s, when the demand for brewers' casks came to an end.

The smaller village cooper could not respond to changing technology and demand in quite the same manner. He would often undertake work for small breweries, but as these were swallowed by larger ones so he lost valuable customers. However, his most important 'raw material' was the old brewer's cask, which, together with wine and spirit casks, could be cut into tubs or knocked down so that the staves could be used to make smaller casks, tubs, buckets or bowls.

The domestic work undertaken by village coopers had been hit earlier than this by the manufacture of cheap galvanised and enamelled buckets, bowls, pots and drums, and the major part of what is known as white coopering, the making of straight-sided vessels, was finished. Commercial directories record the change. In Bedfordshire fifteen coopers were listed in 1853, fourteen in 1861, nine in 1871, three in 1894 and one in 1910. In Hertfordshire there were twenty-eight in 1864, twenty-three in 1878, six in 1899 and in 1902 five.

In the decades during which their numbers were dwindling coopers began to take on other work to survive. The famous sign over the cooper's shop at Hailsham in East Sussex early this century read:

As other people have a sign,
I say—just stop and look at mine!
Here, Wratten, cooper, lives and makes
Ox bows, trug-baskets, and hay-rakes.
Sells shovels, both for flour and corn,
And shauls, and makes a good
   box-churn,
Ladles, dishes, spoons and skimmers,
Trenchers too, for use at dinners.

I make and mend both tub and cask,
And make 'em strong, to make them
    last.
Here's butter prints, and butter scales,
And butter boards, and milking pails.
N'on this my friends may safely rest—
In serving them I'll do my best;
Then all that buy, I'll use them well
Because I make my goods to sell.

It is still possible to come across an old cooper's shop, like the one at Carlton, Bedfordshire, but it is very unusual to see a barrel being made. In Scotland the distilleries still employ coopers, but for the most part they remake whisky casks from American bourbon barrels and in any case new casks would be made with the use of machinery, so that it is very doubtful if many of the coopers still at work would be able to make a cask by hand.

Shaving the inside of a butter churn with a bucket shave on a horse.

Using a saw croze to make the groove in a milk churn.

An eighteenth-century cooperage.

The Barrel

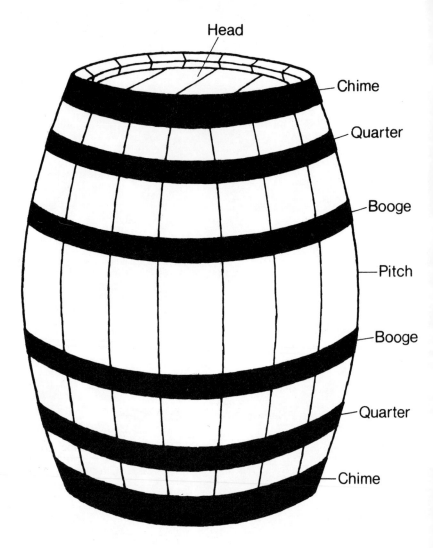

Head

Chime

Quarter

Booge

Pitch

Booge

Quarter

Chime

A section through a Head

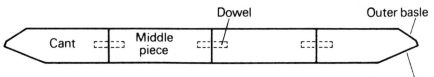

Cant

Middle piece

Dowel

Outer basle

Inner basle

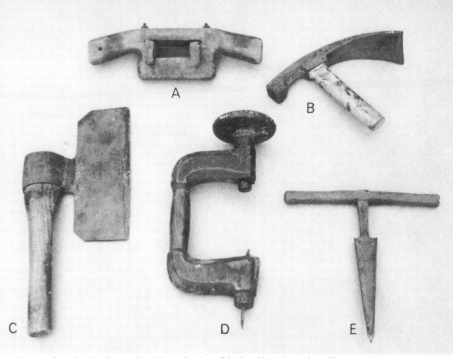

Coopers' tools: A. Buzz; B. Adze; C. Axe (blade offset from handle in order to save hand from being chafed by flying chippings); D. Brace (worked with one hand); E. Auger for tap hole.

MAKING A BARREL

The cooper keeps his tools on his bench or propped up beside it. This is what his bench is for; he works at the block. The tools are kept razor-sharp and tested so that they will cut through a piece of flag or rush with only the force of their own weight, without leaving a ragged edge. They wear so that they sympathise with the wood, and raw linseed oil is rubbed on the working surfaces to reduce friction.

The timber, cut to the appropriate length, width and thickness, is carefully inspected by the cooper for blemishes and to determine which way it will bend more easily without breaking. It is then *dressed*. To do this the cooper holds the stave first across his block and puts a rough shape to it with his axe; this is called *listing* the stave. The stave is then backed and hollowed out with long knives, the longer

staves on the block, the shorter ones on a horse as seen in the photograph (page 15). Lastly it is jointed on a *jointer*, a very long upturned plane, in order to put the *shot* or angle on the edge of the stave, corresponding to the radius of the cask throughout its length. As these staves fit together in the cask the jointing must be so accurate that the joints will be perfect and not allow leakage of liquid under pressure. The cooper also looks carefully down each jointed edge to see how much *belly* he is putting in the cask, which he judges with his practised eye.

When sufficient staves have been dressed they are *raised up* in a *raising-up hoop*, which is filled so as to give the cask the correct capacity; the cooper will judge this to within a pint. It is never wise to put a soft stave next to a tough one, and a good

11

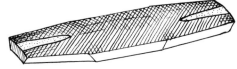

ABOVE: *The stave after being listed.* RIGHT: *The stave after backing.* BELOW: *The stave after being hollowed out.*

ABOVE: *The dressed stave after the last process, the jointing. Dotted lines show the eventual height of the cask.*

The cask after it has been raised up, ready for firing.

A. Horse (in some countries called a mare or a shingle horse); B. The block; C. Block hook.

cooper sorts out his staves very carefully. He hammers the hoop until it is tight and drives a *booge hoop* over the staves; it is now ready for firing.

If the cask is a *stout* one, that is one and half inches or more thick, then the cooper will immerse it in a steam bell for about half an hour or steep it in boiling water to soften the timber and make it more pliable, but if it is *slight,* that is a thin cask, then it is warmed right through by putting it over a *cresset* of burning shavings for about twenty minutes. It is then ready to bend, and with a shout of 'Truss Oh!' two coopers or a man and boy will team up, a stout ash *truss hoop* will be thrown over the cask, still heated over the cresset, and the hammering with trussing adzes will begin. Smaller hoops will be driven down until the open end of the cask

is pinched in. The cask is then turned over. Speed is of the essence—the longer it takes the harder it will be—and you can imagine the exasperation of a cooper, his eyes smarting from the smoke and running sweat, when a hoop is stubborn, or worse, when one breaks under the strain. The truss hoops are driven down on one side until a smaller one can be caught on over the staves, and this process is repeated until the staves are completely bent and the cooper can catch a *dingee hoop,* the same size as the raising-up hoop, on the other end. Care is then taken to keep the hoops on the pitch, that is the belly of the cask, tight, in order to stop any staves from breaking at the weakest point. A cracked stave is called a *duck;* it is an expensive fault. The fire is kept burning in the cresset, watched over by the apprentice,

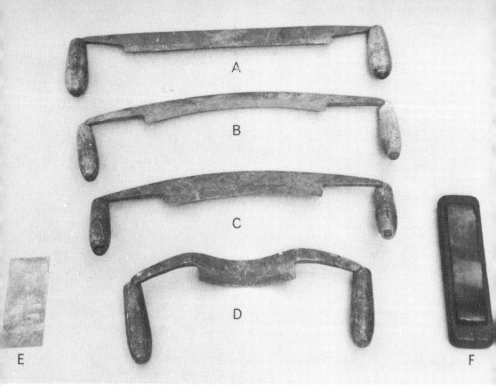

ABOVE: *A. Backing knife; B. Draw-knife; C. Heading knife; D. Hollow knife; E. Scraper; F. Oil stone.*
BELOW: *A. Jointer; B. Beek iron.*

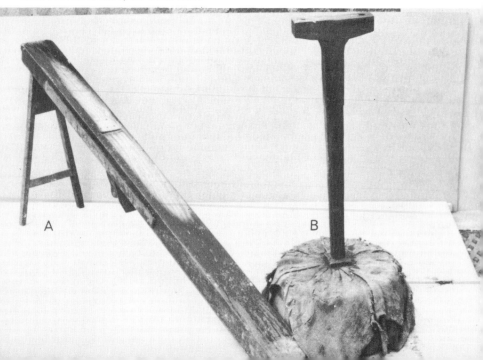

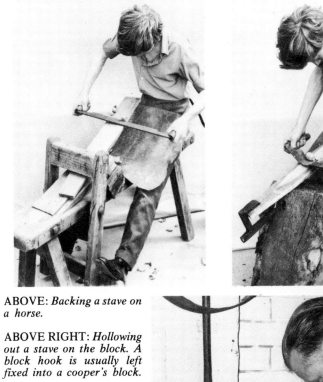

ABOVE: *Backing a stave on a horse.*

ABOVE RIGHT: *Hollowing out a stave on the block. A block hook is usually left fixed into a cooper's block.*

RIGHT: *Jointing the stave.*

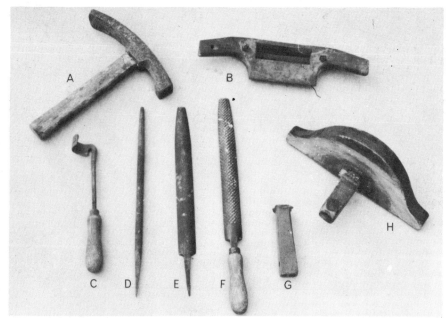

ABOVE: *A. Trussing adze; B. Bucket shave; C. White cooper's round shave; D. Rat-tail file; E. File; F. Rasp; G. Rivet mould; H. Saw croze.*
OPPOSITE: *Listing a stave. Coopers would boast that they could place a silk handkerchief over the block and list a set of staves without catching the silk.*

for at least half an hour after the firing so that the staves acquire *set* and if the hoops are completely removed the staves will retain their shape. At this stage, having been fired, the cask is called a *gun.*

The cask now has to be *chimed,* and a *chiming hoop,* a hoop slightly bigger than the raising-up hoop, is driven on to the cask, replacing the raising-up hoop. The cask is leaned against the block and a bevel or slope is cut on the end of the staves with an adze. To make sure that the chimes or ends are square the cooper then uses a *topping* or *sun* plane around the chime. If necessary he then goes round again with his adze. To make the inside of the cask perfectly curved so that a groove can be cut into it to take the head a cooper uses a *box chiv,* resting the cask between his knee and the block. This can also be done with a *jigger,* a one-handled draw-knife, instead of a chiv, but this requires more skill. A *croze,* the tool which actually cuts the groove, is swung round the inside of the cask in much the same way as the

chiv, care being taken to keep the depth constant. Cutting the groove in a small cask or bucket is done on a horse with a saw croze, keeping the cask or bucket upright. With one end of the cask chimed it can be used as a *case,* the staves being removed in order to repair other casks with them, and some small cooperages bought cases rather than fire their own casks.

The cooper chimes the other end of the cask, but before he cuts the groove he checks the potential capacity of the cask with his *diagonals,* two lengths of metal hinged together, and can adjust the depth of his groove or the way in which the head will fit into the groove so as to adjust the capacity of the cask should he have made a mistake. It only remains then to shave out the inside of the cask with an *inside shave* in order to make sure that it is perfectly smooth so that it can be sterilised effectively and no bacteria will be able to lurk in any crevice. Casks are blistered in the firing and left rough if they are to contain strong beer, wine or spirit which

16

ABOVE LEFT: *Raising up the cask. This is a slight firkin.* ABOVE RIGHT: *Starting to fire a small cask.* BELOW LEFT: *The firing. Driving down on one side in order to catch on a smaller hoop.* BELOW RIGHT: *Blinding smoke and stubborn truss hoops are all part of the process of firing a stout cask.*

The author chiming a kilderkin cask which he has just fired.

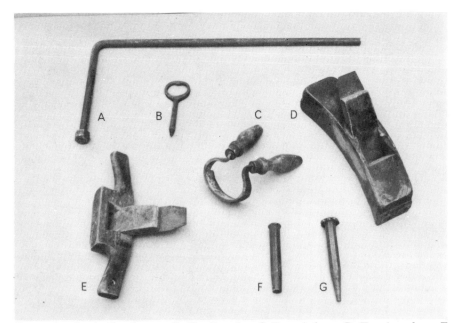

Coopers' tools: A. Knocker-up; B. Heading vice; C. Round shave; D. Topping plane; E. Downright; F. Rivet mould; G. Hoop punch.

will mature in the cask. Coopers bark their arms shaving out casks and the work is never easy.

At this point the cooper usually makes the permanent hoops for the cask. For many casks used in dry work this would mean fitting wooden hoops after the cask heads had been fitted, but in the case of beer, wine or spirit casks, where metal hoops are now used, the cooper buys his hoops cut to appropriate lengths and *splayed,* which is with one edge bruised to fit the bulging shape of the cask. The length of hoop is held round the cask, allowing an inch or so for driving and tension; the cooper holds his thumb on the position where he will rivet and taking the hoop and holding it on the *beek iron* over the hole he hammers a rivet through the iron, through the other flap, and burrs it over with his hammer. The names of the hoops and the parts of a barrel are shown on page 10. Sometimes the cooper will have to bruise more or less splay on to the hoop using the tapered part of his hammer against the beek iron.

The heads are made next, and to find the size required the cooper tries his compasses round the groove, adjusting it until it makes exactly six steps round the groove, which will give him the radius of the head. He then selects his timber and joints the pieces of heading on his jointer. Holding them up against the light he checks his joints and places them together on his *heading board.* After marking the positions of the *dowels* with a piece of chalk he takes his brace and, holding it against his stomach, bores the dowel holes. He makes his own dowels, preferably from pieces of American red oak, and always splits a length of *flag* (rush) and inserts it into the joints of the head before he taps them together. With his compasses he marks the circle and saws it round with a bow saw, leaving a little extra on the *cants* to allow for the head to squeeze in this direction. Bending over the heading board, the cooper then shaves both sides of the head with his *swift.*

Cutting in the head again requires quite considerable skill. It is done with a *heading knife,* holding the head between the body and a notch in the block, the

ABOVE: *Chiming the cask, using an adze to cut the slope on the chime.*
BELOW: *Chiming the cask, using a chiv to level the inside before using a croze.*

ABOVE: *A. Bow saw; B. Driver; C. Hammer (3 lb); D. Chince (for forcing flag into joints); E. Flagging iron (for pulling staves away from the head in order to insert flag).*

LEFT: *A. Heading board; B. Swift; C. Diagonals; D. Cresset (barrel size).*

outer *basle* being cut first, the compass mark again made, and the edge cut exact, with an allowance on the cants. The inner basle is much bigger; the head, in fact, is fitted deeper into the cask than the outside indicates, and sometimes the cooper will use his axe to cut off the surplus wood before he goes round with his heading knife. Having made both his head, top and back, he is ready to fit them into the cask.

To do this he slackens off the hoops on one end and forces his back head inside the cask. Around the groove the cooper inserts a length of flag, puts one side of the head into the groove and carefully turns the cask over so that the head will drop down towards its position. He then taps the head home from the inside, carefully turns it over again and drives the hoops tight on that end. Getting the top head into the groove is more of a problem but the cooper usually uses a *heading vice*, which is screwed into the head in the place where the tap hole will be, and the head is pulled up with this. If he already has a bung hole bored in the centre of a stave then he can put his *knocker-up* through

this and tap the head into position quite easily. It can be done without making any mark on the cask by working a piece of hoop iron between the stave joints.

The outside of the cask is now smoothed with a *downright,* followed by a *buzz,* which is really a turned scraper held in a wooden holder between two handles designed to give the scraper leverage.

The last job is to hammer the hoops home with a hammer and *driver.* On larger casks the cooper will use an eight-pound sledge-hammer, and on the chime hoop a maul. When this is finished the cask will ring like a bell.

When a cooper is making buckets, churns or tubs he employs the same skills in the same processes, but since the vessels are smaller he does many of the operations on a horse.

Although many people call any sort of cask a barrel, in fact a barrel is one which holds 36 gallons. Smaller ones are kilderkins of 18 gallons, firkins of 9 gallons and pins of 4½ gallons. Larger ones are hogsheads of 54 gallons, puncheons of 72 gallons and butts of 108 gallons capacity.

Coopers' tools: A. Croze; B. Inside shave; C. Stoop plane (also for inside shaving); D. Compasses; E. Jigger; F. Chiv (box chiv).

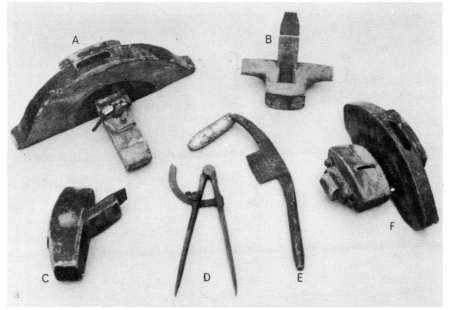

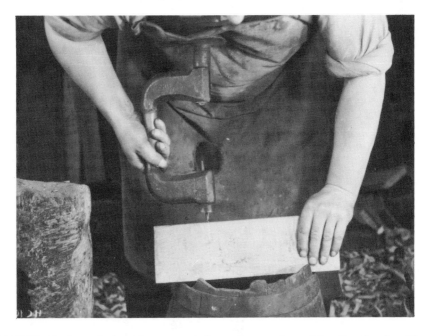

ABOVE: *Boring a hole with a cooper's brace into a piece of heading for dowelling.*
BELOW: *Dowelling the pieces of heading together. Note the flag.*

RIGHT: *Sawing round the head with a bow saw.*

BELOW: *Shaving the head smooth with a swift on a heading board.*

ABOVE: *Cutting in the head, using a heading knife.*
BELOW: *Pulling the top head up into the groove, using an old spike rather than a heading vice.*

LEFT: *Using an inside shave to smooth the inside of a cask.*

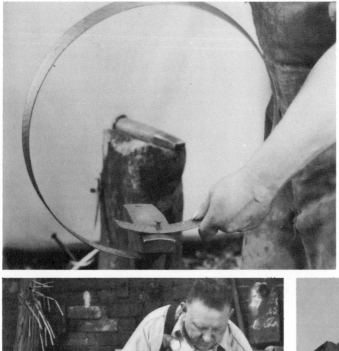

Riveting the hoop.

ABOVE: *A metal driver used centuries ago on wooden hoops.*

LEFT: *Driving a booge hoop on a finished cask using a three-pound hammer and driver.*

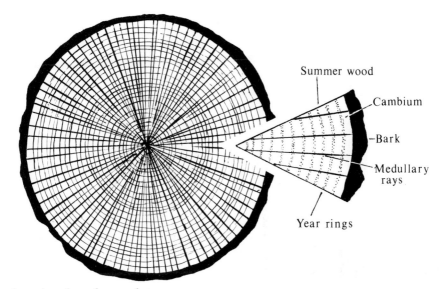

A section through an oak tree.

Labels on figure: Summer wood, Cambium, Bark, Medullary rays, Year rings

TIMBER

For dry coopering the accent was on cheapness and any soft wood or old staves from knocked-down casks would serve for things like fruit, seeds and ironmongery. In wet and white coopering, on the other hand, the kind of wood used and the way in which it has been converted by the foresters is of paramount importance.

The white cooper used mostly Memel oak, some English oak, beech and chestnut for water, milk, butter or cheese vessels where care had to be taken in case the wood imparted a taste to the contents. In most village cooperages you would see high stacks of beer, wine and spirit staves and heading which had been cut into appropriate lengths for making buckets, tubs, churns and casks.

French wine coopers like to use oaks grown in the same earth as the grapes and

wine and spirit casks are always blistered inside during the firing to enable the contents to penetrate deeper into the oak and so aid the maturing process.

Up until the Second World War Memel oak was used almost exclusively for beer casks. It grew along the many rivers flowing down into the Baltic in forests which were predominantly fir, a faster-growing tree, which caused the oaks to be drawn up straight in a fight for light and air, until the 'nurse' firs were cut down, enabling the oak to develop with a near-perfect grain, free from knots, unlike the English oaks, which grow singly and awry.

The Russian peasants, unable to follow their usual occupations in the winter, would cut down the trees while the sap was down, remove the crown (branches) and

A section through an oak tree showing the quartered staves and waste in cleft oak.

bark, cut the trunk into stave lengths, the longest being pipe size, five feet six inches, then split the logs in half with a two-handed axe, or wedges if the log were stouter than two feet six inches in diameter. The log was then split into billets radially and the woodsman marked the timber with a bow string rendered with charred oak, twanged on the oak. The woodsman then worked to this line with his two-handed axe, reducing the log to a parallelogram in section. His skill was such that very little had to be done when the staves were finished off, as they were, on a large plane with one man pushing and one pulling. Staves were shielded from the wind with branches

before being transported to the railhead and thence to the port. Sworn brackers sorted the staves into best quality, crown, and second quality, brack, ploughed with a scribing tool. At three inches thick by six inches wide it was called a full-size stave. They were sold by the mille of 1,200 pieces.

Since Memel oak was split with the grain it was called cleft oak and it was always perfectly quartered, so that the medullary rays radiating from the centre of the tree, which are impervious layers tending to keep their shape during seasoning, run across the staves as in the diagram. The year rings tend to be porous.

English oak cannot be cleft like Memel as the grain is seldom straight enough, and

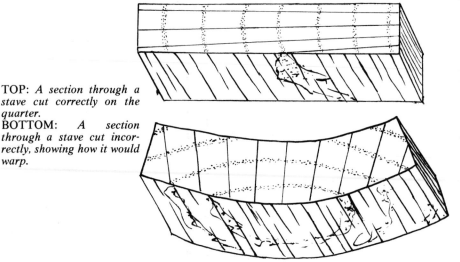

TOP: *A section through a stave cut correctly on the quarter.*
BOTTOM: *A section through a stave cut incorrectly, showing how it would warp.*

it was often sawed so as to give short-grained timber and much that was not well quartered.

Persian oak was used after the Second World War when Memel oak became unobtainable, but this suffered from internal cracking of a honeycomb nature, caused through rapid drying in a hot climate, and twenty-five per cent waste was customary.

American red and white oaks have an excess of acid-tasting tannin and were often lined with a rubber, and later a plastic, solution, but before the First World War Guinness's brewery used to insist on American red-oak casks which had been *pompeyed*, that is charred and blistered inside during the firing process, in order to give the stout its distinctive flavour.

Timber needs to be seasoned, that is dried, traditionally by letting it weather for from two to five years in stacks through which the air could circulate freely. An ordinance put before the Lord Mayor of London in 1488 explains what would happen if unseasoned timber were to be used: 'That where grete deceit and untrowthe dayly . . . by the means of makyng barells . . . of sappy and grene tymber, for lacke of serche and correccion thereuppon to be hadde and done . . . which . . . of necessite must shrynke and . . . lacke of their true and juste measure that they ought to conteyne'.

Preparing a cask for beer or wine would be a problem on which the advice of the cooper would be sought. In order to neutralise the acid-tasting tannin in the oak it is necessary, before filling the cask with beer, to soak it in a solution of salt and sodium carbonate. Casks must have a sweet 'nose', are always 'snifted' before use and, if found to be sour, soaked in a sodium solution. After use a cask should be washed in hot 'liquor' (water) and then, if it is not to be used for a fortnight or more, a quarter of a pint of strong sodium metabisulphite solution should be poured into it, and it should be corked and pegged (sealed), shaken and periodically swilled on the outside to keep the cask from drying out, and the joints from opening up. It can then be left empty for twelve months and will smell very sweet when required, only needing soaking and washing.

ANCILLARY TRADES

HOOP-MAKING

The making of hoops for casks has been a specialist woodland trade for many centuries. It was carried on in Sussex, the Midlands and Furness, and hoops were exported to Jamaica for the sugar barrels, a trade which ended at the turn of this century. They were made mostly of hazel, cut and split with an axe and trimmed with a draw-knife, called a spokeshave. The wood was then steeped in water to make it pliable and coiled on a 'horse', a frame of upright pegs. Coopers would make hoops to size by notching the overlapping ends, binding and nailing them. In Bristol the hooping of casks was in the hands of a separate trade, the hoopers.

Standard sizes of hoops were made by nailing them together within a stout ash hoop. One man could make between four hundred and five hundred hoops a day.

RUSH-GATHERING

Rivermen like Metcalf Arnold, whose yard was on the Ouse at St Ives, now in Cambridgeshire, next to the Ferryboat Inn, used to harvest rushes, wading into the river and cutting them deep near the base. He then spread them out to dry in the sun. The best he called 'coopers' and the rest he sold for mat-making.

Rush, called flag by coopers, swells when it comes into contact with moisture and it was inserted into joints so that if the empty cask were to dry out and the wood shrink, causing the cask to leak when refilled, the rush would swell and 'take up' until the wood swelled.

FURTHER READING

Elkington, G. *The Coopers' Company and Craft.* 1933.
Firth, J. F. *The Coopers' Company.* 1848.
Foster. *A Short History of the Coopers' Company of London.* 1944.
Kilby, K. *The Cooper and his Trade.* 1971.
Salaman, R. A. *Dictionary of Tools Used in Woodworking and Allied Trades.* 1975.

PLACES TO VISIT

Museums with coopers' tools

Ashwell Village Museum, Swan Street, Ashwell, Hertfordshire.
Ashley Countryside Collection, Wembworthy, Chulmleigh, Devon (tel: Ashreigney 226).
Bass Museum of Brewing, Horninglow Street, Burton on Trent, Staffordshire (tel: Burton on Trent 42031).
Bridewell Museum of Local Industries, Bridewell Alley, Norwich, Norfolk (tel: Norwich 22233).
Bygones at Holkham, Holkham-next-to-Sea, Norfolk (tel: Fakenham 710806).
Cookworthy Museum, The Old Grammar School, 108 Fore Street, Kingsbridge, Devon (tel: Kingsbridge 3235).
Curtis Museum, High Street, Alton, Hampshire (tel: Alton 82802).
Guinness Museum, Watling Street, Dublin, Ireland (tel: Dublin (01) 741903).
Luton Museum and Art Gallery, Wardown Park, Luton, Bedfordshire (tel: Luton 36941 or 36942).
Melton Carnegie Museum, Thorpe End, Melton Mowbray, Leicestershire (tel: Melton Mowbray 69946).
Museum of Cider, The Cider Mills, Ryelands Street, Hereford (tel: Hereford 54207).
Museum of English Rural Life, Whiteknights Park, Reading, Berkshire (tel: Reading 85123 ext 475).
Museum of Local Crafts and Industries, Towneley Hall, Burnley, Lancashire (tel: Burnley 24213).
National Museum of Antiquities of Scotland, Queen Street, Edinburgh (tel: 031-556 8921).
St Albans City Museum, Hatfield Road, St Albans, Hertfordshire (tel: St Albans 56679).
Science Museum, Exhibition Road, South Kensington, London (tel: 01-589 6371).
Scolton Manor Museum, Spittal, nr. Haverfordwest, Dyfed (tel: Clarbeston 328).
Watford Museum, 194 High Street, Watford, Hertfordshire (tel: Watford 32297).
Welsh Folk Museum, St Fagans, Cardiff, South Glamorgan (tel: Cardiff 561357).
York Castle Museum, Tower Street, York (tel: York 53611).

Museums with ancient casks

Ashmolean Museum of Art and Archaeology, Beaumont Street, Oxford (tel: Oxford 57522). Anglo-Saxon libation vessels, cooper-made.
British Museum, Great Russell Street, London WC1 (tel: 01-636 1555). Vats excavated at Sutton Hoo; stoup.
Devizes Museum, 41 Long Street, Devizes, Wiltshire (tel: Devizes 2765). The Marlborough vat.
Jewry Wall Museum, St Nicholas Circle, Leicester (tel: Leicester 539111). Mountsorrel bucket (Roman).
Merseyside County Museums, William Brown Street, Liverpool (tel: 051-207 0001). Trawsfynydd tankard (iron age).
Reading Museum and Art Gallery, Blagrave Street, Reading, Berkshire (tel: Reading 55911 ext. 2242).

ACKNOWLEDGEMENTS

Photographs and other illustrations are acknowledged as follows: British Museum, inside back cover; Museum of English Rural Life, University of Reading, pages 9 and 28 (right); St Albans City Museum, pages 15 (bottom), and 17; the Tickenhill Collection, pages 18, 21, 24, 25 (bottom), 26 and 28; Mr M. S. Longuet-Higgins, inside front cover; K. Kilby, all others.
The cover illustration is reproduced by kind permission of the Museum of Cider, Hereford.

OPPOSITE: *A cooper's workshop in the seventeenth century, from a series of engravings by J. J. van Vliet, 1635.*